A Scientific Study of the 2016 White-tail Rut in Minnesota:

The Timing of the Rut

In East MN

By: David Feist

With Contributions from Robert Hannah

Finlayson, MN

Copyright- Nov. 13, 2016

MMXVI

First English paperback edition

TABLE OF CONTENTS

Introduction

Many studies have been done relating moon phases to rut behaviors of male and female white-tailed deer, among those specimens of breeding age in the autumn of the year.

This study was conducted without any use of rut predictor charts from Charles Alsheimer et al. or anything but personally recorded observations made by the study's author, Mr. David Feist, M. Ed.

Only upon the conclusion of this work, then, and only then, will the research conductor consult charts and graphs published in hunting journals regarding the rut forecast for 2016 in the northern latitudes. At that time the publications *Field and Stream* and *Deer and Deer Hunting* will be compared to the findings of these observational data.

In this way, the author's inferences will first occur in a vacuum of deer behavioral expectations.

Annually, many authors describe the time of autumn between November 12th and 15th as peak deer hunting dates in northern latitudes. Even if rut is very active at that time of the year, I am interested in studying daytime movement specifically. Chances are, the movements associated with breeding season may still be occurring at night. (See Data and Conclusion sections.)

"Missouri recently did a study back-dating fetuses from late season harvested does and they proved that over a three year period the peak breeding date (the date when the most does were in estrous) was November 15 plus or minus one day. I always like to hunt during the week that starts ten days before the peak. In this case November 5 – 12. It is tough to beat that time frame."

-**Bill Winke** of MidwestWhitetail.com

Weather data for the end of October and beginning of November for the study has been included, but not in a causal or non-causal regard. In the study's conclusion a section on weather will be mentioned. The reader may interpret the weather data on his or her own, as pertains to two weeks in particular.

Also, in our study area, bucks are almost never observed eleven months of the year. All of the bucks observed in this study are completely out of the ordinary sightings. It can be assumed only that bucks seen here are displaying temporary movements associated with deer breeding season.

Not one buck was seen in a 46-month span on the study property between November 5, 2014 and October 19, 2016. This was most likely due to fawn mortality from the winter of 2013-2014 (after the Fall 2013 deer hunt).

This study and its implications may be considered useful by the Minnesota DNR simply from a data standpoint as there is a Wildlife Management Area adjacent to the study property.

Students in high school and university will notice the use of the scientific method adhered to as much as possible in this work. Please know that if you are interested in a career in wildlife management there are many deer scientists and agricultural colleges which emphasize big game management.

Some university professors and doctors contribute to a scientific monthly publication *Deer and Deer Hunting.* It is good to be emotional and excited about wildlife; however, some other magazines are unrealistic and put far too much pressure on the hunter-enthusiast. It is not practical to always blame oneself for a lack of hunting success when encountering a deer will always involve in some part –a coincidence.

If you consider a career in wildlife management, you might be able to influence policy decisions one day to help the deer herds you care about. In the same way an agricultural biologist can help solve farm animal problems and abuses, you could directly influence deer protection and the solving of ungulate herd health problems the longer you spend in study, preparation, and eventually on-the-job training.

Part 1 - Question

When will the rut behaviors of East Minnesota bucks in Area 156 within zip code 55735 in a one mile area be most evidenced by daytime sightings on white-tailed deer of all sexes and ages, as observed with the naked eye or with oculars or binoculars?

This study is <u>not</u> designed to exactly pattern the different phases of rut. It is designed to draw an inferred correlation between daytime observable movement on deer of both sexes, and the short, annual deer courtship/mating window.

Part 2 – Hypotheses

1. *There will be a build-up of sightings on bucks in the daytime prior to the second new moon of autumn.*

   Sightings will not increase on bucks again then, until after the next full moon. During the new moon, there will be a lull in sightings due to breeding. It is documented that the prey species deer "breeds in the dark so as to birth in the dark" at the end of gestation.

   "Most does are bred under a dark-moon period. Jumping ahead to an average 199-day gestation period, he (LaRoche) calculated that most fawns are born within a day of the third-quarter moon. The adaptive advantage of this is obvious. Fawns born during a dark moon phase have a better chance of avoiding or escaping predators."[1]

   A new cycle of fewer buck sightings will ensue as the primary rut concludes later in November. During the "secondary rut" window the bucks will finish breeding.

2. _Breeding will conclude early this year._

   More bucks will be available to breed because peak rut is occurring prior to gun season. At the start of gun season most of the doe will already have been bred. This early rut will allow some surviving bucks from after the gun season's close (on November 20) to rest and replenish nutritionally in order to survive the winter. Thus, more bucks will survive a potentially difficult winter.

3. _Next year will be an equally good year to see bucks._

   Antlerless hunting in our area is very limited. Therefore, next year's 1 ½ year old buck class is very well protected this year during button fawn stage. [The onset of winter is tardy this year as evidenced by first frost not occurring until later in October, not three weeks earlier in September], and the first two weeks of November will be a good time for young deer to eat heartily in the event there is a tough winter ahead with a lot of snow and/or cold.

   Barring a very snowy winter, there should be an excellent buck season next year as well. Bucks in the 1 and ½ year old class and older are breeding on schedule without needing to service too many extra does [which would deplete their needed energies to survive an expected La Nina winter with extra freezing precipitation.]

Should these snows fly often and become deep, the deer will yard in large groups instinctively. Also, area-wide there are few coyotes or wolves in the study area (if any) to harass local yarded deer as they fend off the elements.

4. *As in the author's past observations, more deer will be seen nearer to the new moon than will be seen in daytime nearest to the full moon.*

This hypothesis is experiential and based on years of good hunts/poor hunts compared to moon phase. The author attributes fewer sightings around the full moon to indicate night rutting and night time feeding. The author asserts that on the new moon half of the month [waning from half-moon toward new and then waxing toward half-moon] there is more daytime movement on deer than there is the other one half of the month. The author asserts this is indicative of: a. rut being nearer to new moon (see above), and b. deer being much more comfortable moving and feeding at *night* on the full moon [and the nights previous and just after full moon] and then they rest more in daytime, accordingly.

This hypothesis is formed from a combined eighty years of hunting experience (myself with accomplished deer hunter Robert Hannah of Knoxville, Tennessee.) Hannah and I have been known for years to cast our lots ahead of the peak-rut seekers and have better successes to show for it. We shun full moon hunting days, disdaining them in any weapons season, whether archery, muzzleloader, or rifle.

Part 3 – Observations and Data Analyses

A. **Deer** Sighted October 20- November 13   **B = buck sighted**

OCTOBER WANING MOON   (rising rut)   **"PEAK"**

18th	20th	21st	22nd	23rd	25th	26th	28th	29th
	13 +1B	8	2	13+2B	11+2B	10+2B	12+1B	1+1B

October's 2nd NEW MOON   (lessening rut)

30th	31st	11.1	2nd	3rd	4th	5th	6th	7th	8th
16	+1B	11	3	6	0	6+1B	6	8	8

[Full Moon- 11/14

9th	10th	11th	12th	13th	[14th
10	2+1B	14	5	1+1B	[1

MN Deer gun season in italics.

[i.e. 11/5-11/20- mostly one buck only w. youth hunt doe tags and a very limited doe lottery]

B. Data

Doe sighted from Oct. 20-Nov. 13, left to right, with bucks in ( )

Peak

2	2	13(1)	8	13(2)	0	11(2)	**10(2)**	0
12(1)	1(1)	16	(1)	11	3	6	4(1)	6
8 8	10	2(1)	14	5	1(1)			

C. Buck Sightings

-five before 10/26

-two on 10/26

-four from 10/27-11/5

-one on 11/10

-one on 11/13

[15]

D. Doe Activity

full moon, then	Oct. 20-25=	47	6 days
PEAK RUT	Oct. 26=	10	1 day
October 27- Nov. 3	10.27-11.3=	49	+8 days
			RUT 17 days
November 5 - Nov. 13, then full moon	11/4-11/13=	60	9 days gun, opened 11/5

(N.B. The 61st deer seen in the gun season was finally a buck on the ninth day of rifle season, because personally I had not seen my son's 8-point live on Nov. 5. Until this buck, I had not seen but one buck since archery-only season, which was occurring before Nov.5.)

E. Average Buck Sightings / Day

October 20 – 25   =   1.0 bucks per day

October 26   =   2.0 bucks sighted

October 27- Nov. 3 =   0.4 bucks per day

Nov. 5 – Nov. 13   =   *0.2 bucks per day*

Nov. 14 – Nov. 20 =   *0.0 bucks per day*

F. Average Doe Sightings / Day

October 20 – 25   =   8.0 doe per day

October 26   =   10.0 doe sighted

October 27- Nov. 3 =   6.0 doe per day

Nov. 5 – Nov. 13   =   *6.7 doe per day*

Nov. 14 – Nov. 20 =   *1.1 doe per day*

G. Camp Deer Taken

October 26th =   seven-point buck

November 5th   =   *(2) eight-point buck/ male fawn*

November 13th   =   *five-point buck*

H. Average Deer Sightings / Day

October 20 – 25   =   8.7 deer per day

October 26   =   12.0 deer day

October 27- Nov. 3 =   6.5 deer per day

Nov. 5 – Nov. 13   =   *7.0 deer per day*

Nov. 14 – 20   =   *1.1 deer per day*

I. Sightings Related to Moon Phase

Toward New	October 20-30	11 days	95
Toward Full	Oct. 31- Nov. 13	14 days	84

J. Daily Average of Sightings

Toward New   October 20-30   8.6 deer per day

Toward Full   Oct. 31- Nov. 13   6.0 deer per day

K. Camp Shots Taken

Buck 1- shot at actual peak of buck sightings (October 26)

Buck 2- shot off-peak when we were seeing 0.2 bucks/day

Buck 3- shot off-peak just before full moon 0.2 bucks/day

Antlerless deer- shot when we were seeing 6.7 doe/day

Part 4 - Conclusions:

Observations indicated that the time of October 20 – November 3, 2016 was excellent for seeing bucks during daytime in a seventeen-day window. The peak day was October 26th. Robert Hannah and I have theorized that the new moon is more significant than the full moon in coincidental increased sightings on deer. However, after the study, the time around the new moon seems to indicate only increased buck movement, with doe movement remaining steady despite moon phase.

The research of Alsheimer, LaRoche, Kenyon theorized rut behavior based on the second full moon of autumn. My research disagrees with theirs. In the Alsheimer, La Roche, Kenyon predictions, a spike in sightings should occur between November 10th to November 23rd. This fourteen day window presumed a very late rut forecasted for 2016.

"According to Alsheimer's Lunar Calendar, major "seeking" behavior should pick up around November 7th and continue until around the 14th, when major "chasing" should begin. This peak in visible rutting activity will continue until around November 21st when the "tending" phase should be kicking into gear and continue through the 28th. According to this prediction, somewhere between November 10th – November 23rd should be some of the best hunting of the entire season." -Mark Kenyon

Therefore our research model and theirs completely conflicts. Based on all of this study's observations, I believe the rut of 2016 to have been early. The Alsheimer team forecasted a much later-than-usual rut. As a team of researchers, they encouraged hunters to plan around their theorized times, again from Nov. 10-23.

As a hunter, I scored on October 26th with bow (buck), my son on November 5th with gun (buck), and a hunter in our party on November 13th with gun (buck). In between the 5th and the 13th of November, 60 doe were seen without a buck anywhere in the main study area in that nine day span. One buck was an exception and was observed on the perimeter of the study area from automobile (11/10).

Weather Impacts

Coincidental to our findings of an earlier 2016 rut, the peak week of rut in October from the Feist/Hannah model was relatively cold. Toward the end of the new moon and the one week of waxing toward full moon there was a time of mostly sunny 61 degree high days seven days straight, and on average low 30s for the morning hunt temperatures. Some days moderated between 40-55, as well. However, there was little cloud cover, some brisk winds over 10 mph, and mostly sunny days, one after another -with extra radiant heat from the sun. The warmest part of these days (daily highs) was around four o'clock in the afternoon.

There was no rain during the warm spell. During the Feist/Hannah model "earlier-rut" observations there was sporadic rain in the cold spell during late October.

The Alsheimer team could not have foreseen a cooler finish to October and a warm start to November, as their prediction was already published in April 2016.

But all weather aside, neither the warm nor the cold changed the total number of doe or deer sightings. Buck sightings happened daytime before gun season, as hypothesized, and some cooler weather coincided with this.

I continue to believe that rut behavior evidenced by increased buck sightings in daytime happens in advance of the second new moon of autumn. These published observations in a 30 day period bear this out in this *short study* done in just a *one breeding year sample,* and only in a *one-mile study area.*

With reference to hypotheses one through four: 1. there was a build-up of sightings on bucks in the daytime prior to the second new moon of autumn, 2. thus, a measure of breeding was likely to have occurred prior to gun season -which did not open until November 5th, 3. the herd *is* poised to be very healthy in the forthcoming 2017 deer season, as well (barring unforeseen climate extremes), and 4. 8.6 deer per day were observed in the anticipated "earlier time" closer to new moon, with 6.0 deer seen nearer to the full moon time of the month, a significant difference of 2.6 deer per day (which is actually 30% fewer deer seen).

One unforeseen result occurred in the data tabulated regarding "average deer sightings per day" unrelated to the moon phases. These results seem to reflect a change in buck sightings mainly, as opposed to sightings of deer overall. This finding is congruent with the frequency of deer sightings year-round in the study area, unrelated to the fall breeding season. However, deer are not seen *daily* year-round in the area, but more so "from time to time," with some patterns of frequency that last longer than is usual.

The number of deer sightings at those times naturally increases, then, with the addition of each year's fawns. Although most of the local doe usually have two fawns in tow, one doe this year did have three.

Before the study, I may have taken the Alsheimer prediction as fact not theory. In the past I have always had an affinity for the *D & DH* researchers and their moon phase ideology. This year; however, I feel they were off considerably.

Another set of predictions was made in *Field and Stream* magazine for November 2016. In an article by Scott Bestul called "The Rut" we found a forecast much more similar to our study's findings. Where our model had proffered the dates October 20th-November 3rd, F&S predicted dates similar to ours at October 25th through November 13th in the Upper Mid-west, the Northeast, and the Mid-west (p. 41).

Interestingly, Bill Winke of *Midwest White-tail* also guessed closer to the East MN rut than full moon phase theorists. Winke asserted that November 5th through November 12th are always good hunting days. Since November 13th was a Sunday; and Minnesota does not have a Sunday hunt ban, and since Bill gave a breeding conclusion "plus or minus one day," *Midwest White-tail* also gave dates corresponding to three-fourths of our hunting successes. Our successes yes, but none of our actual rut observations correspond to Winke's assertion.

Likewise, had we agreed to go by the *Field and Stream* forecast, we still would have harvested all four of our deer. Nevertheless, we would have missed observing 36 doe and three bucks in four days of our very prolific pre-rut observational data. Those three bucks represented the first antlered deer to be seen on the study property in 47 months. Compared to *Field and Stream* then, Area 156 had a much earlier start to the rut, and a noticeably quicker tapering off.

Additionally, all deer sightings in our portion of MN Area 156 ground to a screeching halt after November 13th. (The Alsheimer reports had predicted November 13th to be the fourth day of rut with ten days of good hunting remaining.) We saw five deer in daylight then the week of November 14-18 (the full moon). All other sightings that week occurred in the harvested corn field near the corner of our property after dark, well-within the study area. Twice we saw the same set of fawns in hayfields in daytime without their mother doe around, for a staggeringly low rate then of 1.1 deer per day.

Then an all-out blizzard came mid-day November 18. By then all of our hunting tents and equipment were safely stowed away for the winter, and all our venison was processed, so we wished the die-hards luck the final two days of season November 19-20.

I really hoped a certain ten-year old I knew at a dairy farm nearby would get his first deer then that last weekend. I completely recognized the excitement in his hopes, and wished him that all-important coincidence of sighting a deer in range of his .243 Savage.

Appendices

1. Students in high school will notice the use of the scientific method adhered to as much as possible in this study. In this way, you will see an example of some of the things that go into a science fair project or university study. Thank you for your interest.

2. The second new moon of Autumn 2016 was the new moon of October 30th-31st. There had also been a new moon on October 1st.

3. NOT ONE buck was seen in the study area in the 2015 deer seasons at any time.

4. Not one buck in the 2016 study appeared to be older than 1 ½ years of age.

5. The state hunting lands adjacent were heavily hunted in the poorer 2015 deer season.

6. The state hunting lands were lightly hunted in the 2016 hunting season despite the burgeoned population of bucks and doe in the area.

7. The author believes that after the tough winter of 2013-2014, the deer population in his area quadrupled by Fall 2016.

8. The hunting camp only shot antlerless deer in 2015.

9. The hunting camp only harvested one buck the deer season after the tough winter of 2013-14.

10. The MN DNR's rules on youth hunting are confusing, because of other youth hunting opportunities for deer mentioned in the game regulations pamphlet. We had to ask a lot of questions whether or not there was an early season, separate youth hunt -as in some other states.

11. The hunting camp much prefers harvesting antlerless deer. The author bow hunts, striving to fill his only adult resident tag with a doe before the gun season. This is because of the huge balance differential found in buck to doe ratio. [See Data.]

12. This year the author took a buck in archery, only because the creature was crippled from being hit by a car, and did not have use of its right front leg. And believe it or not, the author's son (on youth tag) harvested the low and small racked eight-point thinking it was actually a doe at 150 yards, twenty minutes after legal shooting hours had begun.

13. This study area could be used to observe for policy-making decisions for Area 156 in the future.

14. The author suggests piloting an "earn-a buck" rule in Area 156. Any hunter in any season then would have to harvest a doe to receive a buck tag -with no party hunting allowed.

15. In terms of wildlife conservation, the author feels a balanced deer herd should outweigh any and all hunters' desires for a large male kill, AT ALL TIMES.

16. Obviously in this study area the 1½ year old males are over-stressed from breeding. The case could be made that such a stressor on an adolescent male deer herd is unethical.

17. Game officials, to keep their credibility as enforcers of strict and detailed laws, should ALWAYS care for the needs of the deer herd first and foremost, and then take care to check for violations. In this way, "the cart will not come before the horse." Kindness to the hunted herd is more important than hunters' feelings or "satisfaction."

18. All baiting ban laws were observed on the hunting property at all times in all seasons between June 2014 and November 13, 2016.

Final Notes:

I wish to thank the DNR of Minnesota for NOT allowing crossbow hunting during the 3.5 month Minnesota archery hunting season. For further information please see the calamitous results of crossbow hunting in maiming/mauling animals which then die from gangrene in swamps. This interview can be located in the book <u>Deer Hunting in North America</u> in its section devoted to expert comments made by archery Pro Shop Owner Larry Flamisch, Jr. of Nazareth, PA.

I would also like to thank my best friend Robert Hannah of Knoxville, TN for taking me deer hunting out of town after one of the absolute worst days of my life.

A message to archers, if you cannot hit a nine-inch paper plate in 10 for 10 shooting prior to archery season, please refrain from hunting with bows. Please note Minnesota's regulations on allowable blades for deer hunting arrowheads, whether using carbon or aluminum arrows. And NEVER use field tips on a deer hunt.

I once heard a buck groan in agony from being hit with a field tip. The hunter's excuse was the field tip "was all he had left in his quiver."

Please care for the deer herd in Minnesota by not putting your wants and desires ahead of the deer's need to be harvested ethically and with excellent marksmanship, always with a mind toward a humane kill.

Please drive warily one hour before and after sunrise and one hour before and after sunset. A deer hit at 55 mph will die. I tell my family of six that those times are the deer's times, not ours, and to be considerate of the deer.

A deer's life is worth more than the car parts used for repairs after you strike them.

[31]

Works Cited

[1]Humphrey, Bob. "By The Light Of The Moon." *Petersen's Bowhunting*. N.p., 28 Oct. 2010. Web. 18 Nov. 2016.

[2]Kenyon, Mark. "2016 Rut Predictions – Could It Be Another Late Whitetail Deer Rut?" *Wired To Hunt*. N.p., 10 May 2016. Web. 18 Nov. 2016.

About the Author

David Feist has authored over 75 books. He can be reached at davidfeistpublishing@yahoo.com.

The Feist Publishing Company's website can easily be located at:

www.easternwoodsbooks.com.